AMANDA ESTERHUYSEN

STERKFONTEIN

EARLY HOMINID SITE IN THE 'CRADLE OF HUMANKIND'

WITS UNIVERSITY PRESS

Wits University Press
1 Jan Smuts Avenue
Johannesburg
2001
South Africa

http://witspress.wits.ac.za

ISBN 10: 1-86814-421-6
ISBN 13: 978-1-86814-421-1

Copyedited by Karen Press
Cover photographs by Sally Gaule
Cover by Limeblue, Johannesburg
Text design by Orchard Publishing, Cape Town
Printed and bound by Creda Communications, Cape Town

CONTENTS

Introduction

The many important fossil discoveries at the Sterkfontein Caves since the 1930s have made this site a household name in South Africa. In 1936 it produced the first adult australopithecine, and in 1947 the most complete cranium of an *Australopithecus africanus* ('Mrs Ples'). A remarkable collection of early stone tools came to light during the 1950s. In 1966 full-time excavations commenced, which in time produced the largest number of individual fossils of a single species of hominid in the world. And most recently, in the 1990s, the discovery of several foot bones led to the unearthing of a near-complete hominid, another world first.

Oddly, although the name Sterkfontein is now synonymous with caves and fossils, it is not known why the caves were given this name. They are situated on the farm Zwartkrans (today called Swartkrans), a name that pre-dates the discovery of the caves. Professor Phillip Tobias, who has researched the modern history of Sterkfontein, surmises that either the farm Zwartkrans once formed part of the larger farm called Sterkfontein, or the caves were named after the Sterkfontein Post Office, which was nearby.

One of the early owners of Zwartkrans was the well-known gold prospector, W.H. Struben, who bought the land in 1884. When the property turned out to be rich in lime and not gold, it was transferred a number of times between mining syndicates, gold estates and individuals. From the late 1880s on, blasting activities revealed caves containing fossils, and the site became a destination for tourists and researchers. In 1945, following the first important fossil hominid discoveries there, the then Historical Monuments Commission proclaimed Sterkfontein Caves a National Monument, and in 1958 the then owners of the land, the Stegmann family, donated the property to the University of the Witwatersrand. At this stage it became known as the Isaac Edwin Stegmann Reserve and was developed by the university as a research facility, devoted to the excavation and study of the caves and its fossils.

In the 1950s the tourist route through the cave and the amphitheatre were developed by the local municipality. The university supported these initiatives and for many years made the caves available to various organisations to run tours and manage public access to the site. In 1998, in recognition of the remarkable fossil record produced by Sterkfontein, Swartkrans and Kromdraai, South Africa nominated the sites and surrounding area for World Heritage status. In 1999 UNESCO inscribed the Cradle of Humankind World Heritage Site, an area of over 47 000 hectares situated in the north-west corner of Gauteng and the adjoining portion of North West Province. At this time the university entered into partnership with the government and made both its land and its intellectual capital available for the development of a small visitor's centre at Sterkfontein and a major exhibition centre at Maropeng.

▲ Bronze bust of Dr Robert Broom at the 'exit' to the Sterkfontein Caves.

The Sterkfontein fossil hominid site is one of several sites presently being excavated in the Cradle of Humankind World Heritage Site.

Sterkfontein has produced more fossil hominids than any other site in the Cradle and, for that matter, any other fossil hominid site in the world. Together with Swartkrans and Kromdraai it formed the basis for the World Heritage nomination. Sterkfontein remains the most accessible site to visitors to the Cradle of Humankind World Heritage Site.

▶ Cooper's

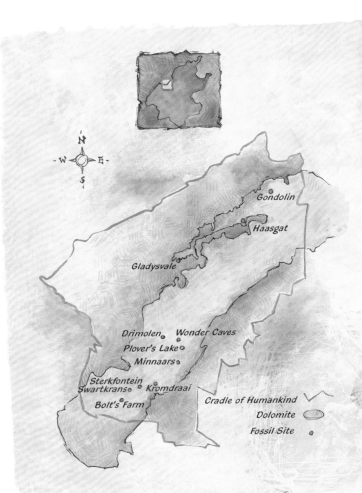

▲ Gladysvale

▼ Sterkfontein

Bolt's farm   Swartkrans

Dolomite, stromatolites and inland seas

2.5 billion years ago

By 2.5 billion years ago, shallow inland seas had formed on our continent's crust. These seas played an important role in the formation of sedimentary rocks like dolomite and chert, and they supported the earliest forms of life.

Formation of dolomite and chert

Waters of seas and oceans are rich in calcium carbonate, the principle ingredient of dolomite. Calcium carbonate is dissolved in seawater, but under certain circumstances it will precipitate out as a solid. In its pure form this solid is called calcite (which makes up limestone), but with the addition of magnesium and iron it becomes dolomite.

We often find the silicate chert embedded in the dolomite. According to one theory, chert precipitates from dissolved silica when fresh water interfaces with salt water. If the inland seas came into contact with fresh water at the time the dolomite was formed, chert could have been produced and embedded in the dolomite.

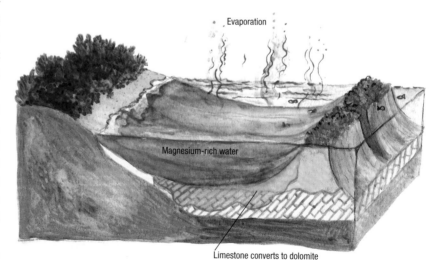

Evaporation

Magnesium-rich water

Limestone converts to dolomite

First signs of life

The ancient sea supported microbial communities made up of different kinds of cyanobacteria. Cyanobacteria are primitive unicellular bacteria (prokaryotes) that contain chlorophyll and are therefore able to photosynthesise. They also contain pigments: red, blue, yellow, green, gold, violet and blue-black. As they share some characteristics with algae they are also referred to as blue-green algae.

Cyanobacteria

Cyanobacteria played an important role in raising the level of oxygen in our atmosphere. The oxygen that we depend on was generated by numerous cyanobacteria billions of years ago. Cyanobacteria were photosynthesising like plants for at least 1.5 billion years before plants evolved, and are thought to be the forerunners of modern day plants.

A mucous layer surrounds most cyanobacteria. Minerals and grains of sediment in the surrounding water get trapped in this sticky mucous layer. As the sedimentary layer builds up, the cyanobacteria move and angle themselves to get light and food, and in doing so begin to create a new sedimentary layer. When these colonies of bacteria bind sediments rich in calcium carbonates they produce stromatolites. There are still stromatolites actively forming today.

What's this got to do with Sterkfontein?

The exposed dolomite in the north-western part of Gauteng has captured and preserved some of the elements of the ancient sea. At Sterkfontein, as you move through the caves that have formed in this dolomite, you will be able to see ripple marks, which formed through the repeated movement of waves as the sea came in and went out, as well as chert and stromatolites.

The stromatolites at Sterkfontein are fossilised. This took place in two ways: either a mineral-rich solution fossilised the bacteria or, more commonly, the bacteria layer disappeared leaving only the sedimentary layers behind.

The creation of a cave

Millions and millions of years have passed. Geological processes have caused mountain ranges to rise to great heights, continents to drift apart, volcanoes to erupt and inland seas to dry up.

The dolomite laid down in the inland sea has become buried under other sedimentary rocks like quartzite and shale.

The Sterkfontein Cave system is about to be formed.

How caves are formed

Groundwater plays a large part in the formation of caves in dolomite rock. It starts off as rain or snow. When it reaches earth, it seeps into the soil and the underlying rocks. Eventually it reaches a level where the ground is saturated with water. The top of this level is called the water table.

Groundwater is slightly acidic because it contains carbon dioxide. Rain and snow dissolve carbon dioxide from the atmosphere as they fall, and, as water seeps through the ground, it absorbs carbon dioxide from decomposing plant material. When this weak carbonic acid seeps into dolomite it begins to carve out caverns by means of dissolution or chemical erosion.

The groundwater works its way through points of weakness in the dolomite, moving down fractures or joints, or along horizontal bedding planes. The carbonic acid reacts with the dolomite and dissolves calcium carbonate or lime out of the rock, slowly creating larger and larger spaces in the dolomite. As the space grows bigger, more water is able to flow through it, introducing physical erosion through water action and increasing chemical erosion through the action of carbonic acid. It takes millions of years for large chambers or caverns to form under the ground.

After a cave has formed, water may continue to drip into the cave. When groundwater saturated with calcium carbonate (lime-charged water) seeps out of a fracture in the roof of a cave it loses some of its carbon dioxide,

> ### Speleothems
>
> Stalagmites, stalactites and flowstones are also called speleothems. Speleothems can tell scientists what the climate was like inside and outside of the cave at the time that each speleothem was forming. And they often trap particles that can be dated by their palaeomagnetic orientation, so scientists can tell how old they are.

and becomes less acidic. At this point it can no longer keep calcium carbonate in solution, and the lime precipitates out and becomes solid. When calcium carbonate is repeatedly deposited on the cave roof, it may form a stalactite or a curtain, among other things. When deposited on the floor of the cave, it may build up in a mound and form a stalagmite, or flow along the floor to form a flowstone. Stalactites and stalagmites often merge to form columns.

12

▲ At present the water-table lies about 40 m below the surface at Sterkfontein. The rise and fall of the water table is monitored at the underground lake.

Spot the rocks and formations at Sterkfontein

Ripple marks

Dolomite

Stalactites

Chert

Stromatolite

Flowstone

14

Tourism affects the caves

The Sterkfontein Caves enable thousands of people to view and understand the processes that produced the remarkable fossil record that resides within it. Unfortunately, visitors have an impact on the sensitive cave environment. Lighting in the caves promotes moss and algae growth and people continue to deface the cave walls by scratching their names into the dolomite.

Narrow shafts and death traps

▲ Shaft-like entrance to the
Sterkfontein Caves

Over many years, the earth and sedimentary rocks that overlay or cover the dolomite erode, so that the dolomite and its underground cave system now lie just beneath the surface of the earth.

Through the continued dissolution of dolomite, a narrow shaft or aven eventually opens between the surface of the earth and the caverns below. This is an important process because, for the first time, things lying on the surface of the earth – like stones, bones, leaves and twigs – can get washed down into the cave below. And creatures may accidentally fall down the shaft and become trapped in the cave, where they die and decompose.

A slope of debris, called a talus, builds up over time. Importantly, at the same time, lime-charged water drips into the mixture and cements the debris together. During this process organic materials become fossilised. This hard mixture of stones, fossils and other infill is called breccia.

16

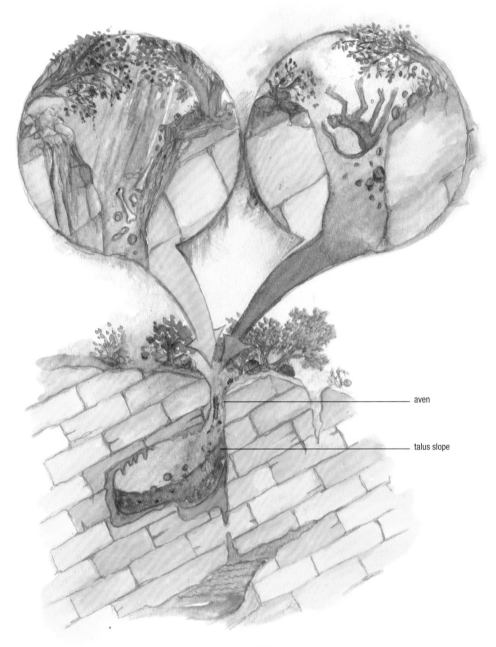

aven

talus slope

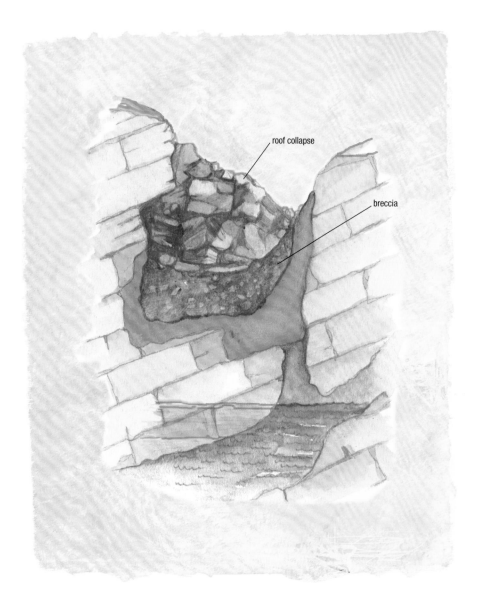

roof collapse

breccia

Over time the roof collapses, sealing the entrance to the shaft and preventing further filling of the cave. Another shaft may then open up, allowing another cave, or another part of the same cave, to fill up and then collapse – and another and another. Each of these caves' infill is like a time capsule, offering a small glimpse into the past through the fossilised material contained within them.

In scientific terms each 'time capsule' is referred to as a 'Member'. At Sterkfontein there are six important Members that provide a window into a crucial period in time during which our early relatives and ancestors were evolving.

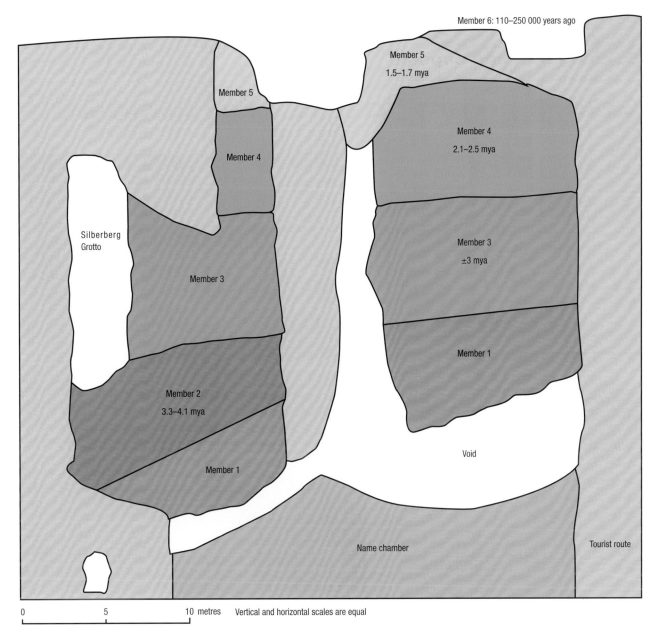

Member 6: 110–250 000 years ago

Member 5
1.5–1.7 mya

Member 5

Member 4

Member 4
2.1–2.5 mya

Silberberg
Grotto

Member 3

Member 3
±3 mya

Member 2
3.3–4.1 mya

Member 1

Member 1

Void

Name chamber

Tourist route

0   5   10 metres   Vertical and horizontal scales are equal

▲ Section through Sterkfontein showing Members and sequence of strata

Gold, lime and fossils

The Sterkfontein Caves remained fairly undisturbed until the discovery of gold on the Reef in the late 1800s. In 1884 the Struben brothers, who were prospecting for gold, bought the farm Zwartkrans on which the Sterkfontein Caves are situated. Although Zwartkrans did not yield gold, it did have another commodity that became important to the gold-mining industry – lime. In 1891 the Macarthur-Forrest cyanide process was implemented to extract gold from ore, and in those days lime was an essential element of this process. In response to the increasing demand for lime, people began quarrying at Zwartkrans.

Around 1894 Guglielmo Martinaglia started a small quarry at Sterkfontein, removing limestone from the side of the koppie. Sometime near the end of 1896 or beginning of 1897, however, he blasted through to the caves below. A number of people began to visit the caves and remarked on the beauty of the stalactites and the brilliance of the crystals and noted the presence of fossil bones. From the start efforts were made to protect the caves, unfortunately much damage was caused by visitors and miners.

▲ Lime kiln in the Makapan Valley. The remains of the kiln that once 'burnt' the lime at Sterkfontein is visible below the excavation site.

Early in the 1930s one of Dart's former students, J.H.S. Gear, visited the site and collected some fossils. Unfortunately, before he could carry out any major study on them, he was granted a post in London. Inspired by a conversation held with Gear around the Wits University Department of Anatomy lunch table, more of Dart's students visited the site and recovered fossils. In 1935 Trevor Jones collected monkey skulls there and published information about them. In 1936 Harding le Riche and G.W.H. Schepers took Robert Broom to Sterkfontein after showing him the monkey fossils. Broom, who was working at the Transvaal Museum, was drawn to Sterkfontein and keen to work there.

Between 1936 and 1939 Broom began recovering and collecting fossils from the lime quarrying activities and made many discoveries, including the lower jaw of a hominid, but he was forced to stop at the outbreak of the Second World War. In 1942 Helmut Silberberg, who intermittently visited and collected fossils from deep in the Sterkfontein Caves, picked up a piece of breccia containing baboon and hyena remains. The hyena, as it turned out, was of utmost importance. It was a primitive form of hunting hyena known as *Chasmoporthetes*, which proved that Sterkfontein was older than anyone had imagined. These new insights prompted Broom to return to Sterkfontein in 1947. There he worked with John Robinson for three years. Dr John Robinson, one of the most important researchers at Sterkfontein, was Broom's assistant from 1945 onward until Broom's death in 1951. Over the years they worked together they recovered, among other things, the near complete cranium of the adult australopithecine now popularly known as 'Mrs Ples'.

In the 1950s Bob Brain and A.B.A. Brink discovered stone artefacts at Sterkfontein. Robinson excavated these and more hominid fossils until 1958. Archaeologist Revil Mason, as well as Mary Leakey, famous for her study of stone tools from Olduvai Gorge in Tanzania, studied the Sterkfontein tools. Kathy Kuman has carried out a more recent study of the ancient stone tools.

By the 1960s the number of hominid fossils recovered from South African sites had placed South African scientists at the centre of many debates around hominid evolution – debates that generated many of the ideas that directed and guided subsequent palaeoanthropological and archaeological research.

In 1966 Phillip Tobias re-opened excavations at Sterkfontein, initially with Alun Hughes and later (from 1991) with Ron Clarke. Phillip Tobias is still an active member of the Institute for Human Evolution, Wits University, and we have his unflagging energy to thank for much of South Africa's contribution to the world of palaeoanthropology.

To date Sterkfontein has produced more than 500 hominid specimens, while animal and plant fossils and stone tools number in the thousands. Fossils are still being uncovered in the caves. In fact, Ron Clarke, with the help of Stephen Motsumi and Nkwane Molefe, made a very exciting find of a near complete skeleton of an australopithecine in the late 1990s.

Plesianthropus transvaalensis

Broom felt that the hominid specimens from Sterkfontein differed from Dart's Taung specimen (*Australopithecus africanus*), and as a result created a new genus and species name for them, *Plesianthropus transvaalensis*. In 1947, after Broom examined the complete cranium and determined it was female, it was nicknamed "Mrs Ples". We now know that the Taung child and Mrs Ples are of the same genus and species, *Australopithecus africanus*. The only difference is that the one is a child (3–4 yrs) and the other an adult.

▲ Sts 5 (Mrs Ples) *Australopithecus africanus*

Who's who?

Prof. Raymond Dart identified the Taung child in 1925 as being an ape intermediate between 'apes and men', or what today we would call a 'missing link'.

Dr Robert Broom, a Scottish physician, took to full-time palaeontology after his retirement. He excavated at Sterkfontein intermittently from 1936 until his death in 1951.

Prof J.T. Robinson worked for many years alongside Robert Broom. Together with Dr Bob Brain he excavated some of the first stone tools discovered at Sterkfontein.

Dr Bob Brain excavated at Sterkfontein through the 1950s. He then went on to excavate at Swartkrans.

Alun Hughes directed excavations at Sterkfontein for just over 20 years.

Dr K. Kuman conducts research on the stone tools found at Sterkfontein.

Dr H.K. Silberberg collected a number of key fossils from the chamber now named after him, the Silberberg Grotto.

Prof. Ron Clarke, famous for his ability to reconstruct hominid fossils, has made one of the most significant hominid discoveries in the world.

Prof. Phillip Tobias's unflagging passion for palaeoanthropology has kept the discipline alive in South Africa.

Sterkfontein is one of a few sites in the world that is excavated on a daily basis by a full-time excavation team. (Left to right) Hendrik Dingizwayo, Isaac Makhele, Nkwane Molefe, Abel Molepolle, Stephen Motsumi, Lucas Sekowe and Solomon Seshwene.

Removing breccia and the fossils contained within it

Fossils are excavated in a number of different ways, depending on how soft or hard the breccia is. Sometimes calcium carbonate, which cements the breccia together, will dissolve and soften the deposit. When this happens the fossils can be dug and sieved out of the loose deposit. If the breccia is still hard, however, blocks of it need to be broken off and the fossils drilled out. Each block of breccia is recorded according to its position in the cave before it is removed. This is done by plotting its co-ordinates into a plan of the site. At Sterkfontein an overhead grid originally provided the coordinates to plot

the material. Today, however, more sophisticated laser technology is often used. Total stations or lazer theodolytes are now used to 'shoot' in the position of fossils, such precision mapping allows one to reconstruct the site and position of each fossil after the material has been removed.

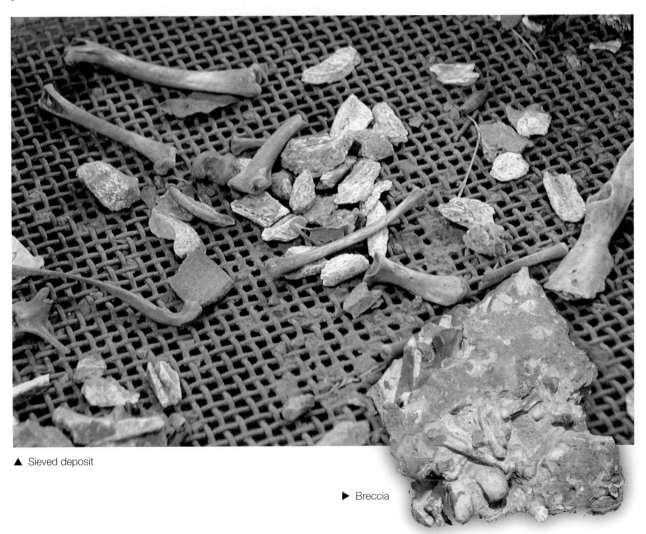

▲ Sieved deposit

▶ Breccia

What is a fossil?

Fossils are the remains or traces of something that lived long ago. Two kinds of fossilisation processes have occurred at Sterkfontein. In the first case, after the soft tissues – like skin, muscles, eyes and internal organs – have decomposed, water that is rich in calcium carbonate and other minerals seeps into the spaces in bones and teeth, and the minerals harden. When some of the bone decomposes, it too is replaced by minerals, which harden in the shape of the bone.

In the second scenario, nothing of the original animal remains, but an imprint or cast of the organ or organism is left behind. It is this process that preserved the 'brain' of the Taung child, which was found in the Taung lime quarry in North West Province and identified by Raymond Dart in 1925. A long time after the child died, the empty skull lay on its side. Sand and lime-charged water washed into the cranium through the hole in the bottom of the skull, filled it up halfway, and in hardening took a print of the inside of the skull. Because our brains leave a print on the inside of our skulls, the fossil cast took on the shape of the brain.

▲ Taung child

▲ Taung brain cast

28

From a 'Little Foot' to a complete skeleton

In 1994 Prof. Ron Clarke's discovery of four foot bones led to one of the most remarkable scientific discoveries of all time.

In 1978 Alun Hughes, then director of excavations at Sterkfontein, started extracting breccia blocks discarded by lime miners from the deeper and older members at Sterkfontein. These Members (1, 2 and 3) are all to be found in what is now called the Silberberg Grotto (named after H. K. Silberberg). After each breccia block was laboriously lifted from the subterranean cave, bones were removed, labelled and stored in boxes.

In 1992 Prof. Phillip Tobias felt that Member 2 warranted further excavation. He, together with Prof. Ron Clarke, organised for some of the breccia, which was particularly hard, to be blasted out from untouched areas in the grotto. By 1994 it was apparent that most of the fossils from this breccia were from carnivores and monkeys. This puzzled Ron Clarke. He wondered why there were no bones from other animals, like buck. So he returned to the collection of bones previously stored away by Alun Hughes. Among foot bones of buck, monkeys and carnivores he found the first of a series of clues that would enable him to make the discovery of the century. He found four bones that belonged to the left ankle and big toe of an early hominid.

Footprints

Footprints, probably of the hominid *Australopithecus afarensis*, are preserved in volcanic ash at Laetoli in Tanzania. About 3.6 million years ago a volcano called Sadiman produced a series of light ash eruptions. The ash contained a mixture of minerals which, when mixed with water or rain, formed a soft cement-like layer. While the layer was still soft, two hominids walked over it leaving their footprints behind. When the layer dried it became rock hard and preserved the footprints. Interestingly, Prof. Ron Clarke, as a young researcher, helped Dr Mary Leakey excavate the footprints at Laetoli. Little did he know that years later he would discover the foot bones of a creature with a very similar foot structure.

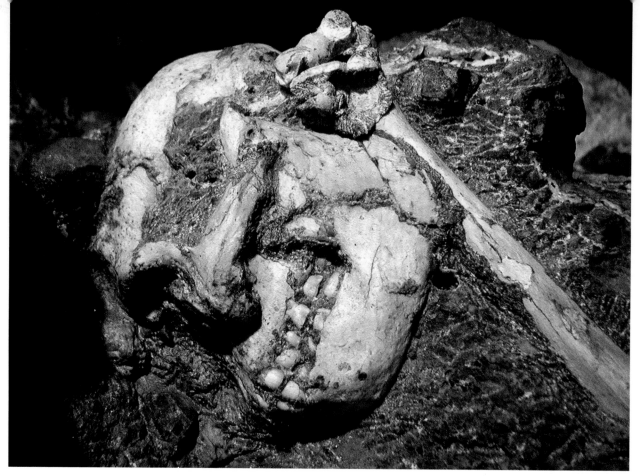

▲ Skull of 'Little Foot' (StW 573)

This find was extremely exciting for two reasons:

1. at the time it represented the oldest hominid find in South Africa; and
2. the fact that these small foot bones had been found together indicated that the creature could have died and decomposed in the caves. If the creature had been killed, then eaten by scavengers and its bones washed into the caves, the chances of the small bones staying together would be slim.

Ron Clarke started going through another box labelled 'monkey fossils' and, relying on his keen memory for shapes and kinds of bones, was able to locate more left and right foot- and ankle-bones, as well as fragments of the left and right shinbones. Now with twelve foot and leg bones of a single hominid, Clarke felt sure there was a chance of finding the entire skeleton. When he examined the right shinbone closely, he realised the bone had been broken when the miners had blasted the breccia out. He made a cast of the end of the bone, and then set what seemed an impossible task for Stephen Motsumi and Nkwane Molefe, two of the fossil preparators at Sterkfontein. He asked them to go down into the enormous, deep and dark cavern to examine all the exposed breccia and find the other end of the shinbone. Mid-way through the second day they spotted a broken section of a bone protruding from the breccia into which their cast fitted perfectly.

In 1997 the three men began to excavate the leg bones in the hope of finding the rest of the skeleton, but beside the lower legs they found nothing more. Knowing that the whole skeleton had to be there, Ron Clarke began to look more closely at the breccia. What he found was that a large block of breccia had collapsed sometime after the creature had died, causing the bottom part of the skeleton to be separated from the top part. Limestone had then flowed in between the two halves. Once this flowstone was removed more bones started to appear – first the left upper arm bone, then the jaw and the skull, and later, the lower left arm and hand.

The creature lay face down with its left arm outstretched above its head. Ron Clarke had made the find of the century, a near complete skeleton of an early hominid.

▲ Legs

▲ Hand   Arm ▼

Common ancestors and family trees

6 million years ago

When scientists began to order and group the animals in the Animal Kingdom, they immediately placed humans in the same category as monkeys, apes and prosimians.

This decision was based on the fact that they have the following things in common:

- well-developed eyes that face forward, providing binocular vision
- flexible fingers and hands that can grasp
- fingers with nails and not claws
- giving birth to one baby at a time, which requires a fair amount of nurturing
- advanced social organisation.

This large group or order was given the name primates, meaning 'first'.

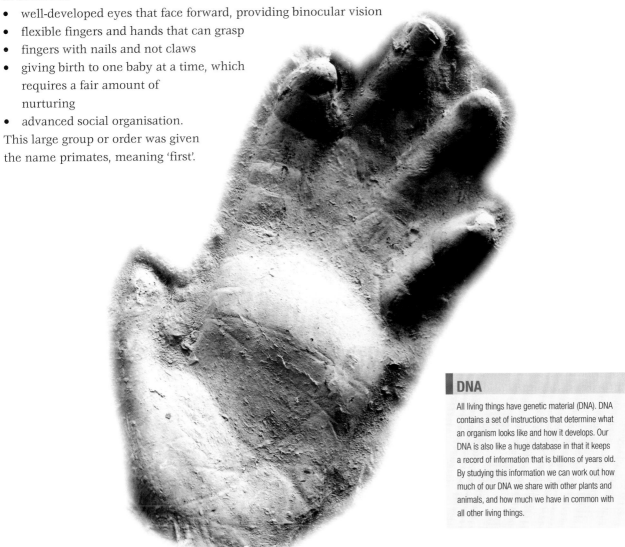

DNA

All living things have genetic material (DNA). DNA contains a set of instructions that determine what an organism looks like and how it develops. Our DNA is also like a huge database in that it keeps a record of information that is billions of years old. By studying this information we can work out how much of our DNA we share with other plants and animals, and how much we have in common with all other living things.

As science advanced it became evident that we were more closely related to some of our fellow primates than previously imagined. Microbiology has shown that we share approximately 98.4% of our genetic material with the chimpanzees and 97.7% with gorillas. In fact DNA studies tell us that we are most closely related to the African apes, with only about 2% of our DNA different from theirs. All these similarities have led scientists to believe that at one time we must have shared a common ancestor with the African apes.

Sometime after 10 million years ago a split between the apes and hominids occurred. The apes evolved in one direction and the hominids in another.

About 6 million years ago, there was a radiation of hominid species and one of these offshoots gave rise to the family of hominids from which we emerged.

Our family

Initially, it was thought that we developed or evolved along a straight line – from ape-like to more human-like, to fully human. We now know, thanks to the many fossil hominid discoveries, that our family tree has many branches and offshoots. In fact we have so many early relatives that it is becoming increasingly difficult to trace our direct ancestry. But, as more and more fossils are found, we are slowly but surely beginning to understand what set our earliest ancestors aside from the apes. For example, evidence from the growing number of hominid fossils in Africa shows that: by 6 million years ago our ancestors walked upright (bipedal), even though they still spent time in the trees; their hands were less specialised than the other apes and therefore more like our own; and analysis of 2.4 million-year-old fossils indicates that some of their diets were more like ours than like the diet of chimpanzees. Indeed, it is only a matter of time before we understand fully how we evolved.

In South Africa the earliest line of hominids that we have are the australo-pithecines, and Sterkfontein in particular has produced a most outstanding record of these early hominids and the environment that they lived in.

The importance of a full skeleton

Around 4 million years ago a long, narrow, vertical shaft opened up between the surface of the Earth and the caves below. Animals, like sabre-toothed cats, hunting hyenas, leopards, baboons and hominids, fell down the shaft, died, decomposed and became fossilised as partial or complete skeletons deep below the surface of the earth.

Fossilised complete skeletons are rare but important finds. They allow the scientists to begin to reconstruct, amongst other things, the length of the animal's limbs, the size of the head in relation to the body, and how powerful

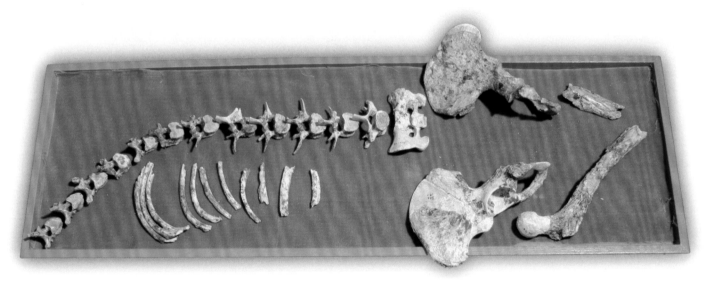

▲ Sts 14

the body would have been. For example, the long forelimbs and skull of the hyena *Chasmoporthetes* tell scientists it ran its prey down in the way that a cheetah does.

As you can imagine, the discovery of an almost complete fossilised skeleton of a 3.3 to 4 million-year-old early hominid at Sterkfontein is even more phenomenal. Scientists from around the world agree that this is the most spectacular find ever. For the first time a scientist, in this case Ron Clarke, will be able to rebuild this early hominid, piece by piece, from its feet to its head.

The near-complete skeleton falls within the genus *Australopithecus* but it has yet to be given a species name. This will only be done once more of the creature has been removed from the breccia and its distinguishing features have been studied. It may be that it is more like the east African hominid *Australopithecus afarensis* (Lucy) or more like our younger species *Australopithecus africanus,* or something completely different that warrants a new species name.

Already this specimen is starting to throw light on what until now have been fairly murky areas of our knowledge of evolution. The arm, hand and legs of the skeleton support arguments that these early hominids were bipedal (they walked upright on two feet), and never knuckle-walked like the apes and monkeys. Clarke explains, 'If you're sitting in the trees you sit upright. When you come down to the ground you have a choice; you can either walk on two legs or run on two legs or you can get down on all fours the way the apes and monkeys do and run on all fours. This hominid walked on two legs. It did not use its hands at all. We can now say this

▲ Giant hyena

for certain because we know it had short arms and short fingers.' In fact one of the things that has surprised the scientific world is that the hand is not like the ape's or monkey's hand, but is about the same as that of humans. Hands of monkeys and apes tend to be quite specialised. They have long fingers and short thumbs and are suited to specific activities. For example, the orang-utan's long fingers enable it to hang from branches. In contrast, humans have a short palm and fingers and a long thumb and are able to carry out a number of different tasks with their hands.

However, the now famous foot bones of 'Little Foot' also tell us that this hominid still spent time in the trees. The big toe was separate from the other toes, a little like our thumb, and enabled it to grasp branches to climb trees. This tree-climbing ability would have helped it escape from the many predators, like hunting hyena and sabre-toothed cats, that we know were around at the time.

▲ Prof. Ron Clarke's reconstruction of 'Little Foot's' foot.

Australopithecus africanus

Australopithecus africanus is by far the most common hominid found at Sterkfontein. There are over 400 specimens of this species, most of which come from Member 4. This does not, however, represent 400 individuals as several fossil specimens, like a mandible, vertebrae, or finger bone could belong to any one individual.

If we compare *Australopithecus africanus* with the chimpanzee (*Pan troglodytes*) and ourselves (*Homo sapiens*) we get an idea of what this australopithecine was like.

The chimpanzee has a muzzle you can place your hand around, like a dog's. It has a very long palate with large canines.

Australopithecus africanus has a protruding face. The shape of its mouth and teeth are more like ours than like those of the chimps.

The human face is fairly flat. It has a u-shaped palate and relatively small canines.

▲ Chimpanzee

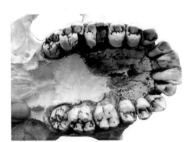

▲ *Australopithecus africanus*

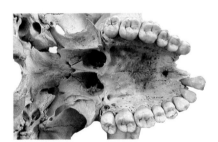

▲ Human

The chimp has large brow-ridges that protect its eyes. It has a low forehead. It has a brain capacity of about 400 cc.

Australopithecus africanus has small brow-ridges, and a low forehead. Its brain capacity is about 470 cc.

▲ Chimpanzee

▲ *Australopithecus africanus*

We have small brow-ridges, and large vertical foreheads. Our brain capacity is between 1200 and 1300 cc.

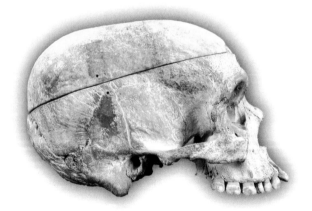

▲ Human

The chimp's hipbone is long and wing-shaped. Its hipbone is shaped like this because it is a knuckle-walker; its upper-body weight is distributed between its shoulders and its hips.

Australopithecus africanus's hips are very similar to ours, demonstrating that by 2.4 million years ago its hips were fully adapted to carrying upper-body weight.

▲ Chimpanzee

▲ *Australopithecus africanus*

We are fully bipedal. Our hipbones are compact and slightly s-shaped, designed to carry upper-body weight, and in the female to give birth to large-brained infants.

▲ Human

Inner ear and balance

Deep inside the inner ear we find the organ of balance. The organ of balance helps synchronise our vision with our head and body movements. When the organ of balance malfunctions we feel dizzy and sea-sick. Scientist Dr Fred Spoor has discovered that there is a link between the size of the semi-circular canals and locomotor behaviour, and, more importantly, that the semi-circular canals fossilise. Studies of the organ of balance in *Australopithecus* show that it is like that of the great apes, indicating that it combined climbing behaviour with bipedality.

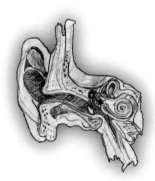

Size

Australopithecus africanus as an adult was not much bigger than a 9-year-old human – approximately 1 m in height.

Diet

It was previously thought that the diet of the australopithecine was very much like that of the modern chimpanzee, which predominantly eats fruit. Chemical analysis of the teeth of *Australopithecus africanus* indicates that they were more generalist. The chemical signature suggests that they enjoyed a mixed diet that included an equal spread of meat, insects, tubers, fruit and seeds.

▲ Prof. Ron Clarke addressing a group of people who are seated on what remains of Member 4

The environment

Australopithecus africanus was not at the top of the food chain, like you and me. We know from the nature of the fossil remains that various large cats and other predators ate these creatures. In fact the large amount of hominid and animal bones found in Member 4 probably washed into the caves after they had been devoured and picked clean by predators and scavengers.

Member 4 is quite unique in that it contained over 300 pieces of fossilised wood in addition to thousands of animal specimens. The study of these fossils has enabled scientists to reconstruct the environment in which *Australopithecus africanus* lived.

Fossil wood preserves the detailed cell structure of the original specimen, even though it has technically turned to stone. Under the microscope you can see all the cellular details of the fossil wood, as well as the arrangement of the different cells and vessels, which differs from plant to plant. By comparing the cellular structure of fossil wood with that of modern wood, you can identify the wood to genus and often species level.

Some of the specimens from Sterkfontein were identified as *Dichapetalum*, which come from a large genus of plants of mostly lianas or woody vines. The species identified are only found today in warmer, more tropical areas of central and western Africa. This information, together with the knowledge that lianas do not have fibres to support themselves – they climb over trees using them as support to reach the sunlight – tells us that 2.4 million years ago there was a much wetter climate in the region than there is today, and the area around the caves was more wooded. This wooded area would have provided a home for the now extinct giant colobus monkey, also found in the fossil record.

▲ Colobus monkeys

The fossils of other animals like buck, rats and shrews indicate that there was also open grassland nearby. This means that a forest probably ran along the river, but became open woodland with patches of grassland futher away from the river.

◀ Artist's reconstruction of the Sterkfontein environment 2.4 million years ago.

Chalicothere

The *Chalicothere* had a large heavy body like that of a rhinoceros, but a head like that of a horse. It had three toes with claws; these were possibly used to tear vegetation down from trees.

Sabre-toothed cats

Three different sabre-toothed cats have been excavated from sites in and around Sterkfontein. All three were in the area by 2.8 million years ago but were no longer found shortly after 1.5 million years ago. *Megantereon* was a large, powerful cat. It was about the size of a large leopard but built as solidly as a lion. This cat had long, elongated canines that it probably used to slash at its prey. *Homotherium*, a lion-sized sabre-toothed cat, had shorter canines that were serrated like a bread knife. *Dinofelis* had shorter canines and a more compact powerful build. While this animal was probably not a fast runner, it seems its powerful forelimbs would have enabled it to hold and subdue its prey.

▲ *Dinofelis*

▲ *Megantereon-Canine*

46

Leopards

The leopard is one cat that has existed in its present form for millions of years. Its success has been attributed to the fact that it feeds on a wide range of different mammals, from small mice to large antelope.

Dassies

Two kinds of dassie (*Hyrax*) have been identified from the Sterkfontein fossil collection. The first, *Procavia antiqua*, is smaller than the modern common dassie, and the second, *Procavia transvaalensis,* was one-and-a-half times bigger than the modern dassie.

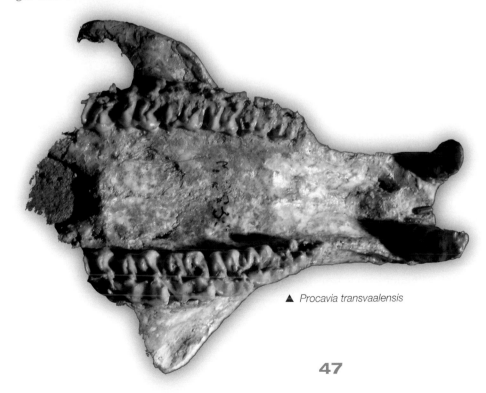

▲ *Procavia transvaalensis*

The flat-faced hominid

As more and more discoveries are made and research techniques improve, it is increasingly evident that in the past there was more than one hominid in the landscape at any one time. Some went extinct and others evolved. One, of course, gave rise to our own genus and species, *Homo sapiens* – presently the only remaining hominid in the landscape.

By two million years ago early *Homo*, as well as another hominid known as *Paranthropus robustus*, roamed the Sterkfontein Valley. *Paranthropus,* although not that common at Sterkfontein, is extremely common at other sites in the area.

Contrary to what you would think, the name 'robustus' does not apply to the overall size of this hominid, but rather to the size of its teeth and jaws. *Paranthropus* had a very large, open, flat face, and massive flat teeth – even its canines were flat. Its bottom jaw was also very large and heavy. All these features were adaptations to the food that it ate. The large flat teeth and heavy jaw enabled it to grind hard gritty foods, like grass seeds, nuts, berries and the like, in order to digest them.

Paranthropus needed very powerful muscles to grind and crush things between its jaws. In very large adults these powerful muscles caused them to develop a crest on top of their heads. This crest formed because, as the jawbone and associated muscles grew bigger and heavier, the muscles running from the lower jaw under the cheekbones to the top of the head required more bone to which they could anchor.

There are enough complete skulls of *Paranthropus robustus* to be able to distinguish between males and females. Recently, Dr Andre Keyser discovered a near-complete skull of a female at the site of Drimolen, just north of Sterkfontein. This female, nicknamed Eurydice, does not have a crest but she has the flat face and characteristic teeth of a *Paranthropus*.

Paranthropus went extinct about 1 million years ago. In total its kind roamed Africa for over 1.5 million years. There are various theories explaining

Eurydice

Dr Keyser called the female *Paranthropus* Eurydice, because it was lying near a much larger male jaw. This reminded him of the mythological character Orpheus who went to the underworld to find his bride Eurydice. Drimolen has also produced several juvenile specimens. One of the babies was only about eight months old at the time of death.

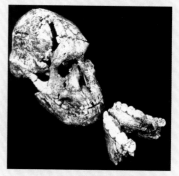

why *Paranthropus* went extinct. They range from large-scale environmental changes that reduced its food base, to its inability to compete with the rapidly evolving *Homo*.

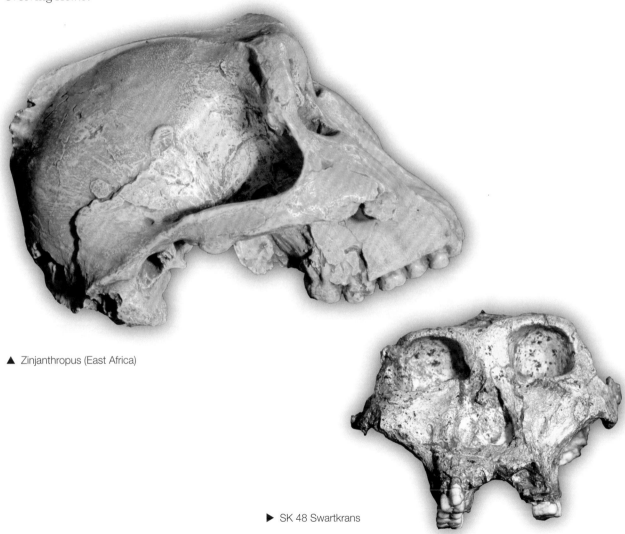

▲ Zinjanthropus (East Africa)

▶ SK 48 Swartkrans

Early technology and *Homo ergaster*

Our early ancestors or relatives first started making and using stone tools just over two-and-a-half million years ago. The earliest stone tools are called Oldowan tools after Olduvai Gorge in Tanzania, where they were first found. Oldowan tools found at Sterkfontein date to between 2 and 1.7 million years ago.

Most Oldowan tools are made in one of two ways. The first method, called bipolar technique, involves taking a rock – quartz in the case of Sterkfontein – placing it between two stones and then hammering down with the top stone. As quartz is brittle, it cracks and pieces flake off. These flakes are very sharp and can be used to butcher and to cut meat and plants for food or other

▲ Core and flakes

StW 53 – a case for cannibalism?

In 1976 Alun Hughes, then field director of excavations at Sterkfontein, discovered some teeth and cranial fragments in a sinkhole filled with decalcified breccia. Over a period of four days he and Ron Clarke excavated over thirty pieces of the skull and then found part of the same skull in the wall of the sinkhole. Once reconstructed this specimen became known as StW 53. This specimen, thought to date to between 2 and 2.5 million years ago, has cut marks on its cheekbone. Dr Travis Pickering, who studied the marks on the bone surface, maintains the cut marks were inflicted by a stone tool, and that the pattern of cut marks is identical to those found on animals that have been butchered. This means that a stone tool was used to de-flesh StW53. Was the creature eaten by a fellow hominid of the same species, or by a more distant and possibly more advanced relative who lived at the same time?

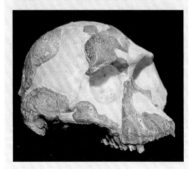

uses. The other method that is more common is a freehand hard hammer technique, whereby one rock is hit with another. This once again produces sharp flakes.

There is some debate as to who made these early tools. There are at least two hominids in the landscape around Sterkfontein who could have been responsible for the tools. The first is an early form of *Homo* or its immediate ancestor, and the second is *Paranthropus robustus*, a hominid that went extinct. It has been argued that *Paranthropus* would have been capable of making the tools and, furthermore, that *Paranthropus* remains have been found in deposits containing tools. However, it has also been argued that because the technology does not disappear when *Paranthropus* goes extinct, either all hominids were making these tools or the only toolmaker was *Homo*, whose line develops this ability through to the present.

Over time another kind of hominid evolved called *Homo ergaster*. It is this hominid that we associate with the more specialised tools that begin to appear at Sterkfontein around 1.5 million years ago. These tools are intentionally shaped to carry out specific jobs. For example, these hominids shaped hand axes – large pointed tools – for hacking and bashing, to remove limbs from animals and remove marrow from bone. They also made cleavers with sharp, flat cutting edges to carry out more heavy-duty butchering. These specialised tools are called Acheulean tools – named after the French site, St Acheul.

Interestingly, even though the tools are named after a French site, they only appear in Europe after about 500 000 years ago. This tells us that the technology originated in Africa and spread to Europe and Asia through the movement of hominids out of Africa.

Homo ergaster was a much more modern kind of hominid. It had a larger brain size, a more modern face and is generally bigger than anything we have seen before. In fact it has a body very like our own. What is interesting is that *Homo ergaster* appears to be ranging over larger areas of territory. We know this because we find large accumulations of flaked and unflaked rock at

Homo ergaster

Homo erectus is by and large only found in Europe and Asia while *Homo ergaster* predominates in East and South Africa. Some researchers believe that *Homo ergaster* is really the same species as *Homo erectus*, which eventually gave rise to *Homo sapiens*. Others, however, think *Homo erectus* was a distinct species that did not give rise to *Homo sapiens*.

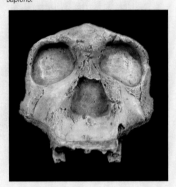

various sites around the country. *Homo ergaster* thus appears to be making use of a wider range of habitats, both wet and dry. At Sterkfontein Acheulean tools and some fossils of *Homo ergaster* have been found in Member 5.

▲ Cleaver

▲ Handaxe

The site of Swartkrans, across the road from Sterkfontein, is the site best known for *Homo ergaster*. It is also from this site that we get one of the earliest indications that hominids may have controlled fire. Scientist Bob Brain found a consistent build-up of burnt bone in the Member 3 deposit (around 1 million years ago) at Swartkrans. Every square that he excavated in Member 3 had burnt bone and almost every level had burnt bone. In other words there was a 6-metre build up of deposit filled with burnt bone.

Dr Brain suggests that the reason for a sudden influx of burnt bone is that hominids sitting near the entrance of the cave were tending fires. The burnt bone washed into the cave later. He feels it is unlikely that they were able to make fire, but rather that they captured and controlled natural grass fires. A study of the bones showed that they had been burnt at a very high temperature for a long time. As bones burnt in an incidental veld fire are not burnt to this degree, this lends further support to his idea that the bones were burnt in a tended fire.

Bob Brain also identified over 60 bone tools from Swartkrans. He hypothesised that the bone tools were used to dig out roots and tubers from the ground. More recently, these bone tools were studied by Dr Lucinda Backwell, who found that striation marks that she detected on the ancient bone tools closely matched those on replicated tools that she had used to dig into termite mounds. She argues that the early hominids ate termites because they provide an excellent source of energy. Termites have a very high fat and protein content; in fact they have a higher calorific value than both fish and steak.

▲ Worked bone

Member 5   Member 4

▲ Member 5 dates to between 1.5 and 1.7 million years ago and is associated with *Homo ergaster* and Acheulean technology.

Out of Africa

All the earliest evidence for hominids is found in Africa. Sometime after 2 million years ago the first of our own genus, *Homo,* started to move beyond the bounds of Africa. Initially, they explored and colonised the warmer reaches of Europe and Asia, and then as they became more adept at surviving in more adverse conditions they were able to penetrate cooler regions.

As in Africa, there was a radiation of species in Eurasia, and so we hear about *Homo ergaster, Homo erectus, Homo antecessor, Homo heidelbergensis, Homo neanderthalensis* and later, only about 35 000 years ago, about an early form of modern human, Cro-magnon. There is still much debate about which of the earlier species gave rise to modern humans (*Homo sapiens*). Part of the debate arises from the fact that modern humans appear to have evolved in Africa between 120 000 and 160 000 years ago, long before they appeared in Eurasia. Scientists thus argue that there could well have been a second migration out of Africa, during which the modern humans from Africa colonised the rest of the world, displacing Neanderthals and other species. However, modern human remains have recently been discovered in China that appear to date to around 130 000 years ago. This may imply that *Homo sapiens* evolved from ancestral stock in Africa, Asia and Europe.

Neanderthals

Neanderthals occur in Europe and Asia between 200 000 and 30 000 years ago. At one stage Neanderthals were depicted as intellectually advanced, showing signs of burying their dead, and were thought to have inter-bred with the ancestors of modern humans. However, many of these burials have been contested and recent DNA analyses have caused people to question their relationship to modern humans.

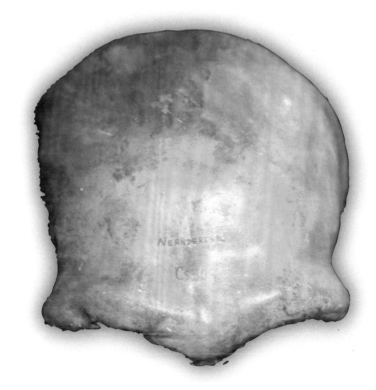

▶ *Homo neanderthalensis*

How are these sites and fossils dated?

A number of different methods are used to date geological members and fossils. It is important to use more than one independent means of dating to cross-check your results. In other words, when many different methods point to one date, the scientist can be confident that the date is probably correct.

The first step usually involves sequencing the members. This is done by determining the order in which the members formed and by examining the types of fossils found in each member. For example, we know three-toed horses preceded one-toed horses, therefore a member containing three-toed

▲ A flowstone sample removed for dating

horses must be older than one containing a one-toed variety. This is called 'relative dating', since we don't know an exact date but we know how old the layers are relative to each other.

The second step is to compare fossil animals from Sterkfontein and other local sites with those from East African sites. In East Africa, sites are well-dated because there are many layers of volcanic material, which have been dated very accurately by radiometric methods such as the Argon-Argon method. Some of the animals found in South African sites are similar to those in East Africa, so we can get a rough estimate of when the animals existed. This is a comparative dating technique.

More recently Prof. Tim Partridge has applied a range of new techniques to date the 'Little Foot' skeleton. One of the methods is called palaeomagnetic dating. Prof. Partridge explains that this technique involves measuring reversals in the Earth's magnetic field. Every so often the North Pole changes its position to that of the South Pole. This flip happens quite rapidly and has occurred many times in the past. At the moment we are in a normal magnetic period but 780 000 years ago the magnetic field was reversed. These reversals are well understood through the radiometric dating of lava, which retains the magnetic signature of the time when it erupted. This has given us a worldwide palaeomagnetic time scale. At Sterkfontein this technique was used to date the new complete skeleton found in Member 2. Scientists found that when stalagmitic flowstone built up in Member 2, small quantities of dust containing iron oxides became trapped in the flowstone, and these iron oxide particles preserved the magnetic signature of the time. By sampling the flowstone from below and above the skeleton they discovered five separate magnetic reversals; when these reversals were matched up with the world-wide palaeomagnetic time scale it provided a date of 3.33 million years for the skeleton.

Another technique that has been applied involves measuring quantities of radioactive aluminium and beryllium in quartz crystals found in the cave sediments. This was carried out with the help of scientists from Purdue in Indiana, USA, and provided a date of between 3.5 and 4.1 million years for 'Little Foot'.

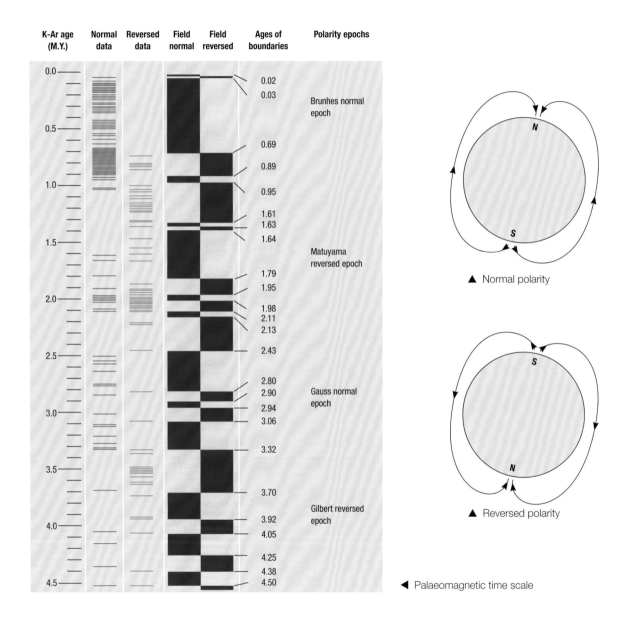

K-Ar age (M.Y.)	Normal data	Reversed data	Field normal	Field reversed	Ages of boundaries	Polarity epochs
0.0					0.02	
					0.03	Brunhes normal epoch
0.5						
					0.69	
					0.89	
1.0					0.95	
					1.61	
					1.63	
1.5					1.64	
						Matuyama reversed epoch
					1.79	
2.0					1.95	
					1.98	
					2.11	
					2.13	
2.5					2.43	
					2.80	
					2.90	Gauss normal epoch
3.0					2.94	
					3.06	
					3.32	
3.5					3.70	
						Gilbert reversed epoch
4.0					3.92	
					4.05	
					4.25	
					4.38	
4.5					4.50	

▲ Normal polarity

▲ Reversed polarity

◀ Palaeomagnetic time scale

Conclusion

South Africa has undertaken to protect, conserve and promote the fossil sites on the Cradle of Humankind World Heritage Site. This involves eliminating or minimising the natural and human factors that threaten to damage and devalue the site.

Sites with public access, like Sterkfontein, are often most at risk and need to be carefully monitored. Infrastructure like lighting, stairways and railings impact on the micro-environment within the cave and alter the habitat of creatures that live within the caves. All caves, for example, have a twilight and dark zone. The introduction of lighting into the dark zone encourages the growth of algae and moss and has a negative impact on those mammals and invertebrates that exist in darkness. The added dust and carbon dioxide that accompanies tourist groups also affects the growth of stalactites, stalagmites and other formations. Problems are also encountered when water that runs off the cement stairways and hardened pathways causes important cave infill to erode.

Tourists are, of course, not the only cause for concern. Pollution of the underground water caused by industry and farming poses a major threat to any dolomitic cave system. It is for this reason that the water in the underground lake is tested on a regular basis. The caves are also continuously inspected for structural changes that may be introduced by natural forces like flooding, earth tremors and even tree roots.

The real value of Sterkfontein lies in its research potential, and for this reason the effects of research on new and old material are also carefully considered. Researchers' proposals for new studies at the site are weighed up against the damage the proposed analysis may cause to the fossils, and newer and less invasive techniques are continually developed for this purpose. Excavation is by its nature destructive, and because even the sediments around the fossils contain valuable information, new and more sophisticated excavation techniques are sought to ensure that the maximum amount of information is recorded before the fossiliferous material is removed.

With careful and continued management, the Sterkfontein Caves will continue to be a popular destination for local and international visitors. There is no doubt that the site and fossils will continue to produce valuable information about the early hominid and other animal populations that roamed the landscape millions of years ago. The excavation of new areas may even take us further back in time and clarify issues about our ancestry. Certainly, Sterkfontein and its researchers will remain central to debates about early hominid evolution for many years to come.

Further reading

Blatt H., Middleton G. & Murray R. 1980. *Origin of Sedimentary Rocks,* second edition. New Jersey: Prentice Hall.

Clarke, R.J. 1998. First ever discovery of a well-preserved skull and associated skeleton of Australopithecus. *South African Journal of Science* 94: 460–463

Clarke, R.J. 1999. Discovery of complete arm and hand of the 3.3 million-year-old Australopithecus skeleton from Sterkfontein. *South African Journal of Science* 95: 477–480

Keyser A.W. 2000. The Drimolen skull: the most complete australopithecine cranium and mandible to date. *South African Journal of Science* 96: 189–197

Kuman, K. & Clarke, R.J. 2000. Stratigraphy, artefact industries and hominid associations for Sterkfontein, Member 5. *Journal of Human Evolution* 38: 827–847

Marean, C.W. 1989. Sabertooth cats and their relevance for early hominid diet and evolution. *Journal of Human Evolution* 18: 559–582

Partridge, T.C., Shaw, J., Heslop, D. & Clarke, R.J. 1999. The new hominid skeleton from Sterkfontein, South Africa: age and preliminary assessment. *Journal of Quaternary Science* 14 (4): 293–298

Pickering T.R., White, T.D. & Toth, N. 2000. Brief Communication: Cutmarks on a Plio-Pleistocene Hominid from Sterkfontein, South Africa. *American Journal of Physical Anthropology,* 111: 579-584

Spoor, F. 2000. Balance and Brains: Evolution of the human cranial base. www.leakeyfoundation.org

Spoor, F., Wood, B. & Zonneveld, F. 1994. Implications of early hominid morphology for evolution of human bipedal locomotion. *Nature* 369: 645–648

Thackeray, F. 2000. 'Mrs Ples' from Sterkfontein: small male or large female? *The South African Archaeological Bulletin* 55:155–158

Tobias P.V. 1973. A New Chapter in the History of the Sterkfontein Early Hominid Site. *Journal of South African Biological Society* 14: 30–44

Tobias P.V. 1979. The Silberberg Grotto, Sterkfontein, Transvaal, and its Importance in Palaeoanthropological Researches. *South African Journal of Science* 75: 161–164

Tobias P.V. 1983. The Sterkfontein Caves and the role of the Martinaglia family. *Adler Museum Bulletin. Festschrift in honour of Dr Cyril Adler on the occasion of his 80th birthday*: 46–52

Popular Books

Blundell, G. (ed). 2006. *Origins: The story of the emergence of humans and humanity in Africa*. Cape Town: Double Story.

Bonner, P., Esterhuysen, E. & Jenkins, T. (forthcoming). *A Search for Origins: Science, history and South Africa's 'Cradle of Humankind'.* Johannesburg, Wits University Press.

Johanson D. & Edgar, B. 1996. *From Lucy to Language*. Johannesburg: Wits University Press.

McCarthy, T. & Rubridge, B. 2005. *The Story of Earth & Life: A southern African perspective on a 4.6-billion-year journey*. Cape Town: Struik Publishers.

Visitors Centres

Sterkfontein Caves Tel: (011) 668 3200

Maropeng Tel: (011) 577 9000, Website www.maropeng.co.za

Transvaal Museum Tel: (012) 322 7632, Website: www.nfi.co.za

Popular Websites

www.talkorigins.org

www.indiana.edu/~origins

www.becominghuman.org

www.cradleofhumankind.co.za

Acknowledgements

Prof. Ron Clarke and Dr Kathy Kuman who tirelessly promote and support Sterkfontein and educational initiatives at the site. They, together with Prof. Phillip Tobias, Professor Emerites at the University of the Witwatersrand, were instrumental in obtaining funding for the first edition of this booklet from the Ford Foundation.

Prof. Phillip Tobias, Prof. Tim Partridge, Dr Bob Brain, Dr Francis Thackeray and Dr André Keyser for taking time away from research to provide information for the public.

Susan Buss and Dr Jeanette Smith helped develop the Archaeological Resource Development Project educational initiative at the Sterkfontein Caves.

Susan Smuts for editing the first edition.

Photo credits

The following people kindly contributed artwork for the book:
Sally Gaule's photographs are on the front and back cover and pages 5, 7, 13, 14, 15, 16, 21, 22, 26, 35, 36, 42, 49, 54, 55, and 64. Amanda Esterhuysen's photographs are on the front and back cover and pages 14, 20, 21, 23, 26, 27, 28, 29, 30, 31, 37, 38, 39, 40, 41, 43, 44, 46, 47, 50, 51, 52, 53, and 56. Christine Andrews' artwork is on pages 7, 8, 12, 17, 18, and 42.

The following organizations kindly granted permission to use various artwork in this publication: Archaeological Resource Development Project, Cradle of Humankind World Heritage Site Archives, Transvaal Museum, Northern Flagship Institution and Faculty of Health Sciences, University of the Witwatersrand.

▲ Fossil tray